Visualizing Linear Algebra and Differential Equations
with Diagrams and Flowcharts

The Guide to Seeing the Bigger Picture

Megan Lim

Geometric understanding is just as important as conceptual understanding. The ability to visualize minimizes the need to memorize.

The goal of this book is to establish a greater comprehension of the relationships within Linear Algebra and Differential Equations through diagrams, flowcharts, and illustrations. The stepping back to look at the bigger picture can greatly help when later zooming in to scrutinize the smaller details. While at first there may not seem to be any obvious connection between Linear Algebra and Differential Equations, hopefully by the end of this book, the relationship between these two concepts will have made itself more apparent.

A Preview:
Understanding the Relationship Between Linear Algebra and Differential Equations

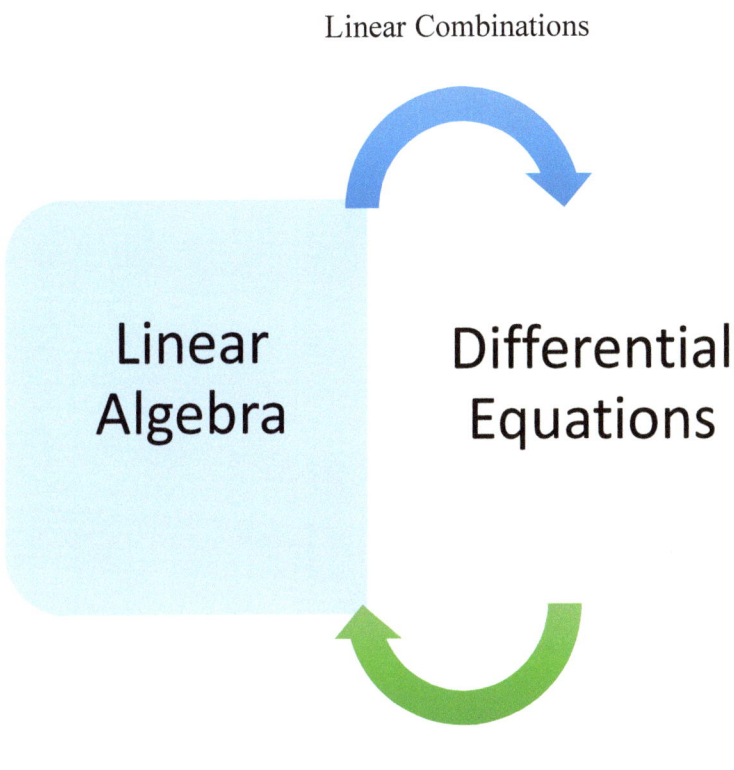

the idea of Linear Combinations links Linear Algebra and Differential Equations

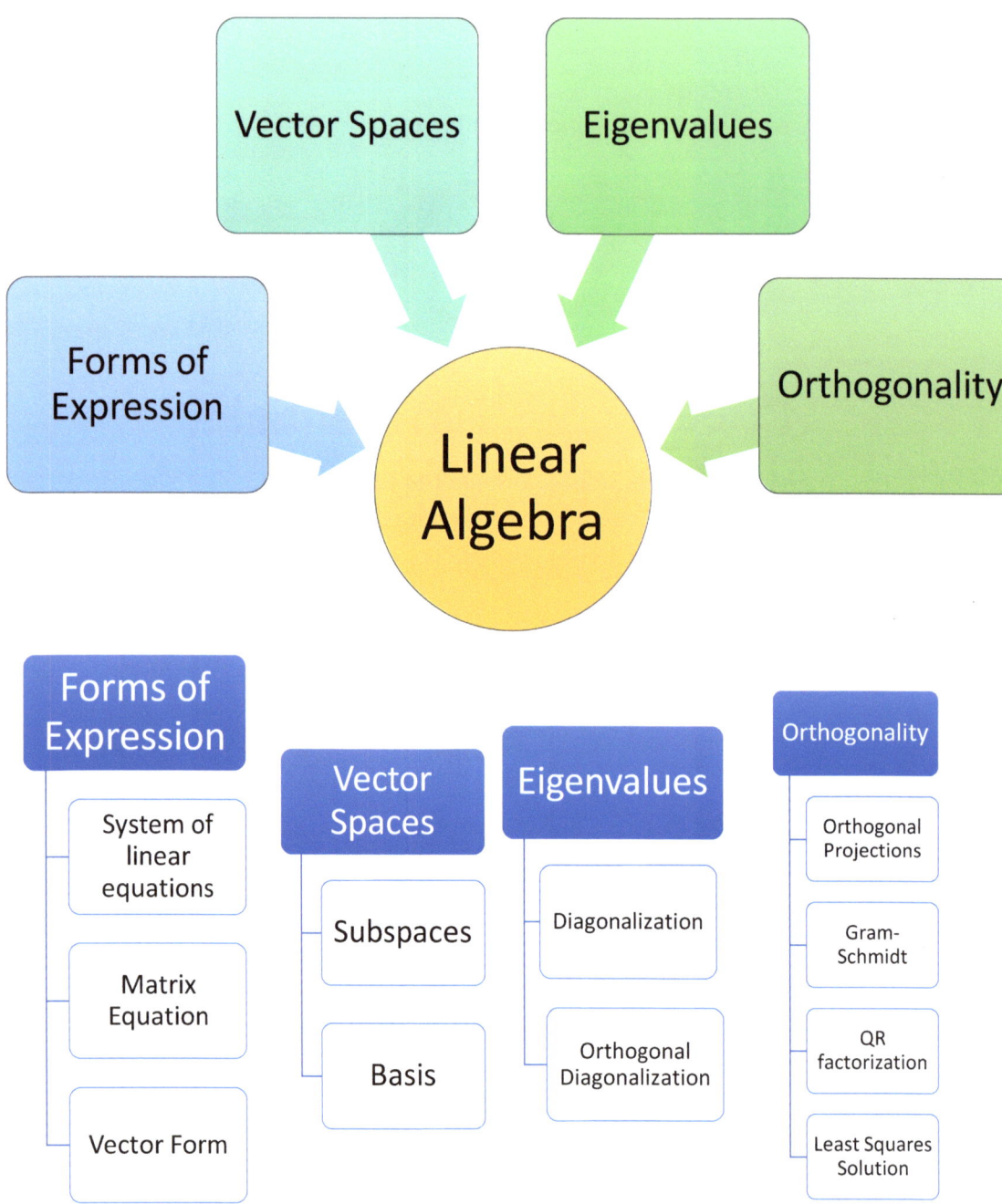

System of Linear Equations

$$a_1x_1 + a_2x_2 + \ldots + a_nx_n = b$$
$a_1 \ldots a_n$ — scalars/weights
$x_1 \ldots x_n$ — variables, solution inputs

Vector Form

$$a_1x_1 + a_2x_2 + \ldots + a_nx_n = b$$
$a_1 \ldots a_n$ — scalars/weights
$x_1 \ldots x_n$ — vectors

$$x_1 \begin{bmatrix} a_{11} \\ a_{21} \end{bmatrix} + x_2 \begin{bmatrix} a_{12} \\ a_{22} \end{bmatrix} + x_3 \begin{bmatrix} a_{13} \\ a_{23} \end{bmatrix} = \begin{bmatrix} b_1 \\ b_2 \end{bmatrix}$$

Matrix Equation

$$Ax = b$$

b is a linear combination of columns of A using corresponding entries in x as weights
matrix A acts on x to transform it into b

$$\begin{bmatrix} a_{11} & a_{12} & a_{13} \\ a_{21} & a_{22} & a_{23} \end{bmatrix} \begin{bmatrix} x_1 \\ x_2 \\ x_3 \end{bmatrix} = \begin{bmatrix} b_1 \\ b_2 \end{bmatrix}$$

Geometric Interpretation: set of all linear combinations of $x_1 \ldots x_n$ is a span

Span as a line {v}:

v

Span as a plane {u,v}:

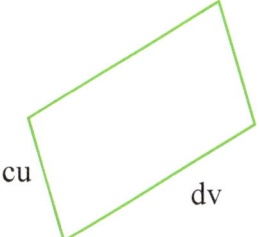
cu   dv

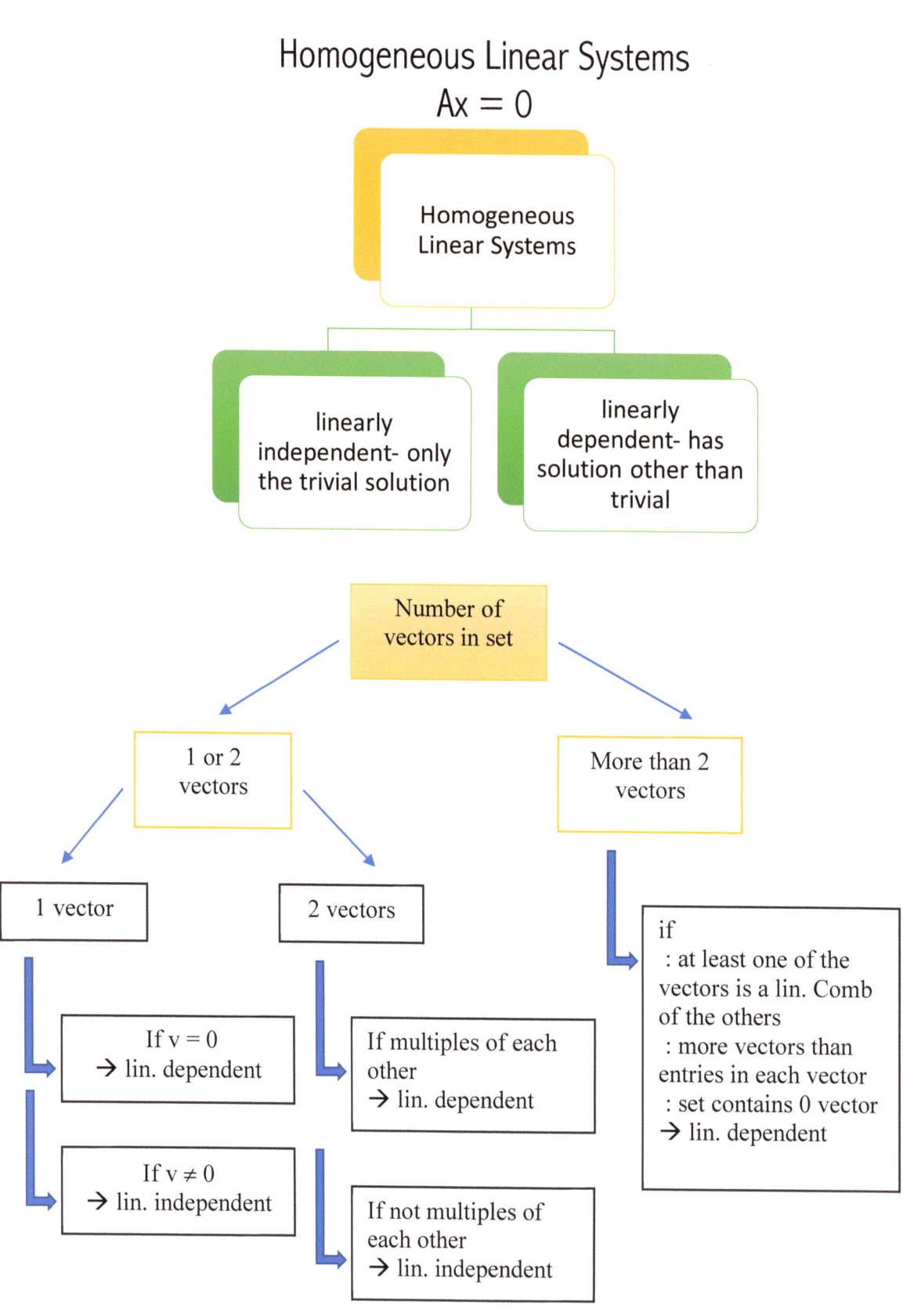

Matrices in Linear Algebra

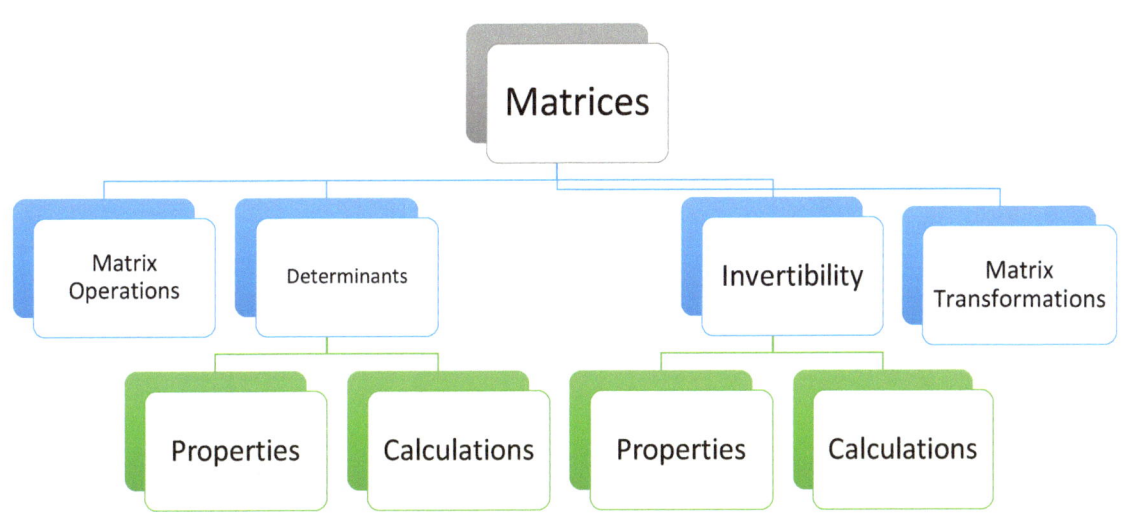

Matrix Operations

Sums and Scalar Multiples
- $A + B = B + A$
- $(A + B) + C = A + (B + C)$
- $A + 0 = A$
- $r(A + B) = rA + rB$
- $(r + s)A = rA + sA$
- $r(sA) = (rs)A$

Multiplication
- $A(BC) = (AB)C$
- $A(B + C) = AB + AC$
- $(B + C)A = BA + CA$
- $r(AB) = (rA)B = A(rB)$
- $IA = A = AI$

Transpose
- $(A^T)^T = A$
- $(A + B)^T = A^T + B^T$
- $(rA)^T = rA^T$
- $(AB)^T = B^T A^T$

Some Warnings:
Usually, $AB \neq BA$
If $AB = BA$, it is not necessarily true that $B = C$
If $AB = 0$, it is not necessarily true that $A = 0$ or $B = 0$

Determinants
2x2 matrix- area of parallelogram
3x3 matrix- volume of parallelepiped

Properties
- If A is a triangular matrix, then det A is product of entries along main diagonal
- det A^T = det A
- det (AB) = (det A)(det B)

Row Operations (A = square matrix)

If a multiple of one row of A is added to another row to produce a matrix B		then det B = det A
If two rows of A are interchanged to produce B		then det B = -det A
If one row of A is multiplied by k to produce B		then det B = $k \cdot$ det A

Calculations

2x2

$A = \begin{bmatrix} a & b \\ c & d \end{bmatrix}$  det A = ad - bc

3x3

$A = \begin{bmatrix} a & b & c \\ d & e & f \\ g & h & i \end{bmatrix}$  det A = a(ei - fh) - b(di - fg) + c(dh - eg)

Invertible Matrices
(n x n matrices)

- columns of A are linearly independent
- linear transformation is one to one
- has only the trivial solution
- maps $R^n \rightarrow R^n$
- has n pivot positions
- **Invertibility**
- $AA^{-1} = I$ and $A^{-1}A = I$
- row equivalent to identity matrix (I)
- $\det A \neq 0$
- A^T is invertible

Finding Invertible Matrix

Row Reduction
$[A \mid I] \sim [I \mid A^{-1}]$

Unique solution
$Ax = b$
$x = A^{-1}b$

Formula

2x2:
$$A^{-1} = \frac{1}{ad-bc} \begin{bmatrix} d & -b \\ -c & a \end{bmatrix}$$

3x3:
$$A^{-1} = \frac{1}{\det A} \, adj\, A$$

adj A:
1. find matrix of minors
2. checkerboard of signs
3. transpose

Properties
$(A^{-1})^{-1} = A$
$(AB)^{-1} = B^{-1}A^{-1}$
$(A^T)^{-1} = (A^{-1})^T$

Matrix Transformations

Every matrix transformation is a linear transformation.
Every linear transformation is not necessarily a matrix transformation.

matrix transformation- type of transformation
linearity - property of a transformation

T: transformation: function: mapping

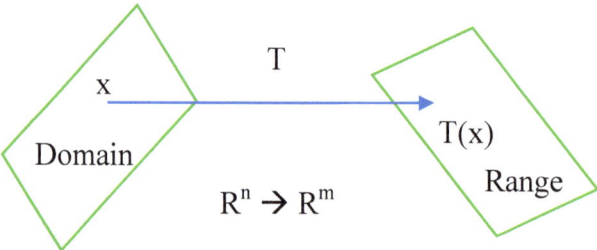

T: **onto**: for each b in R^m there is *at least one* x in R^n

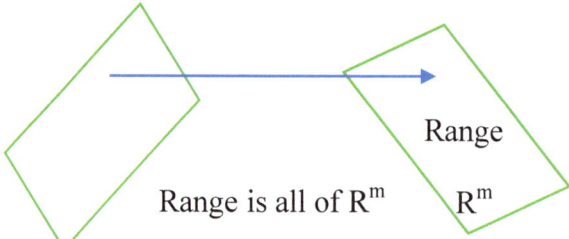

T: **one-to-one**: for each b in R^m there is *only one* x in R^n

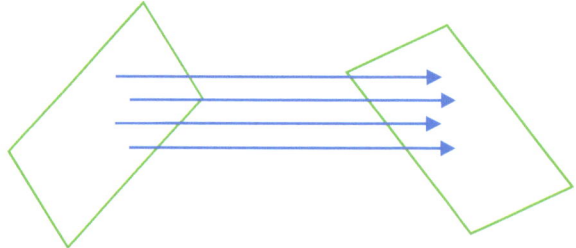

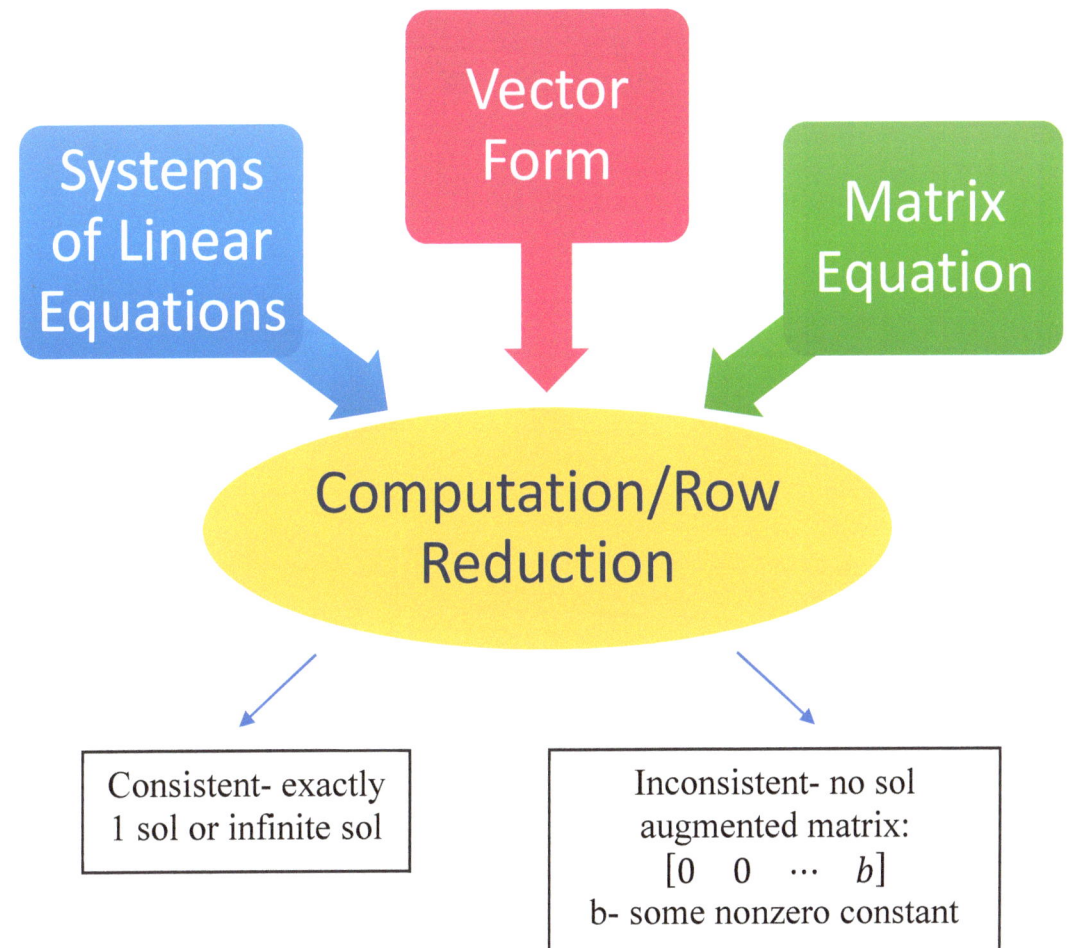

Vector Spaces

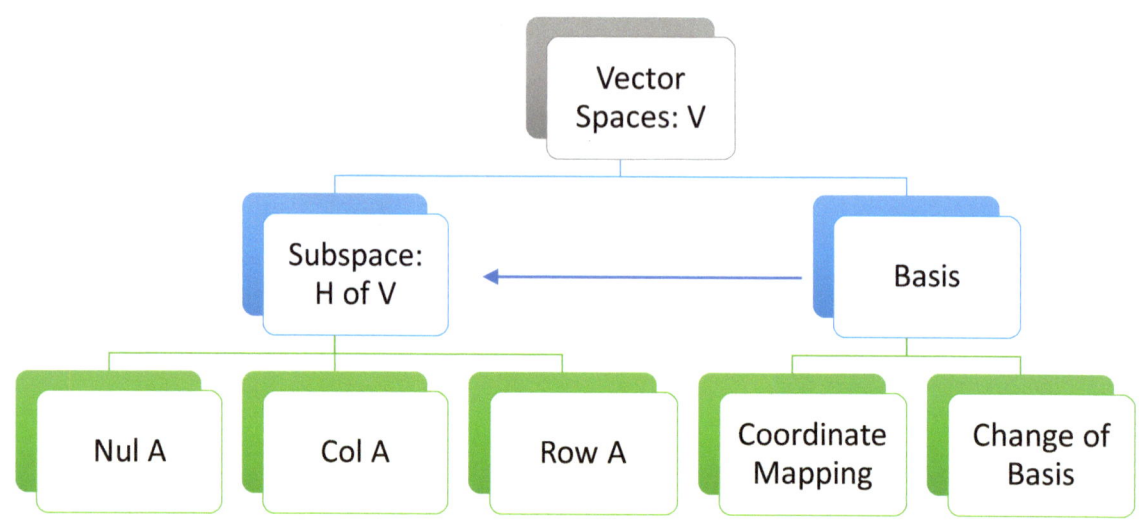

Vector Space Properties
- $u + v$ is in V
- $u + v = v + u$
- $(u + v) + w = u + (v + w)$
- a zero vector exists in V such that $u + 0 = u$
- for each u in V, there is a vector $-u$ in V such that $u + (-u) = 0$
- scalar multiple of u: cu is in V
- $c(u + v) = cu + cv$
- $(c + d) = cu + du$
- $c(du) = (cd)u$
- $1u = u$

Subspace Properties
*linear combination of vectors = span = subspace
- zero vector of V is in H
- closed under vector addition: $u + v$ is in H
- closed under multiplication by scalars: cu is in H

Col A (range)	Nul A (kernel)	Row A
• set of all linear combinations of columns of A • subspace of R^m • basis: pivot columns of A • dim = # pivot columns	• set of all solutions of homogeneous equation • subspace of R^n • basis: expressing in row reduced echelon form • dim = # nonpivot columns (free variables)	• set of all linear combination of rows of A • subspace of R^n • basis: zonzero rows of echelon form B

$$\text{rank } A = \dim \text{Col } A = \dim \text{Row } A = \text{rank } A^T$$

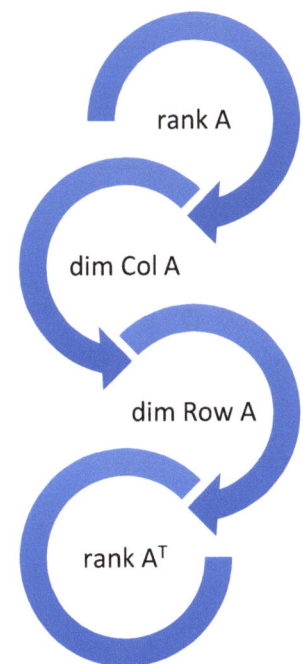

Basis

$$\mathcal{B} = \{b_1, \ldots, b_p\}$$

the most efficient spanning set

Conditions:
1. $\mathcal{B}$ is a linearly independent set
2. $\mathcal{B}$ spans H

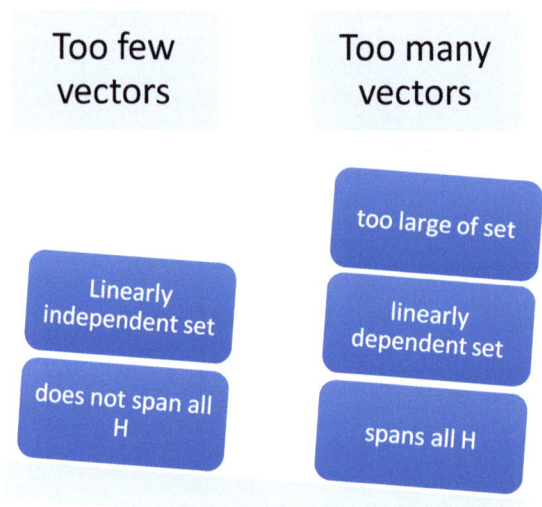

Coordinate Mapping

one-to-one linear transformation

Basis: $\mathcal{B} = \{b_1, \ldots, b_p\}$

coordinate vector of x: weights on the basis:

$$[x]_\mathcal{B} = \begin{bmatrix} c_1 \\ \vdots \\ c_n \end{bmatrix}$$

$$x = \mathcal{B}[x]_\mathcal{B}$$
$$\mathcal{B}^{-1}x = [x]_\mathcal{B}$$

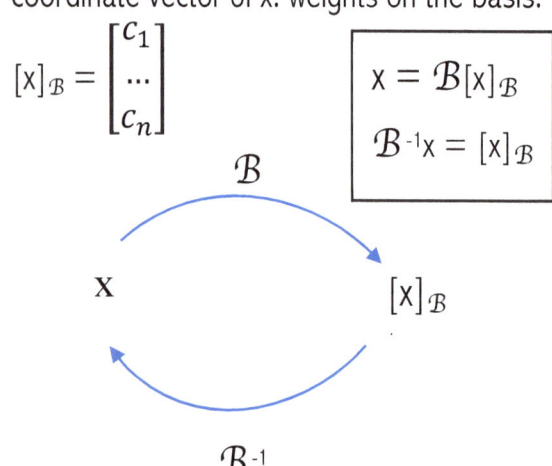

Change of Basis

changing basis from $\mathcal{B} \rightarrow C$

$$[x]_C = \underset{C \leftarrow \mathcal{B}}{P} [x]_\mathcal{B}$$

$$[c_1 \quad c_2 \ | \ b_1 \quad b_2] \sim [I \ | \ \underset{C \leftarrow \mathcal{B}}{P}]$$

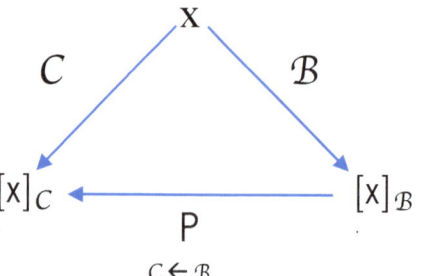

Eigenvalues

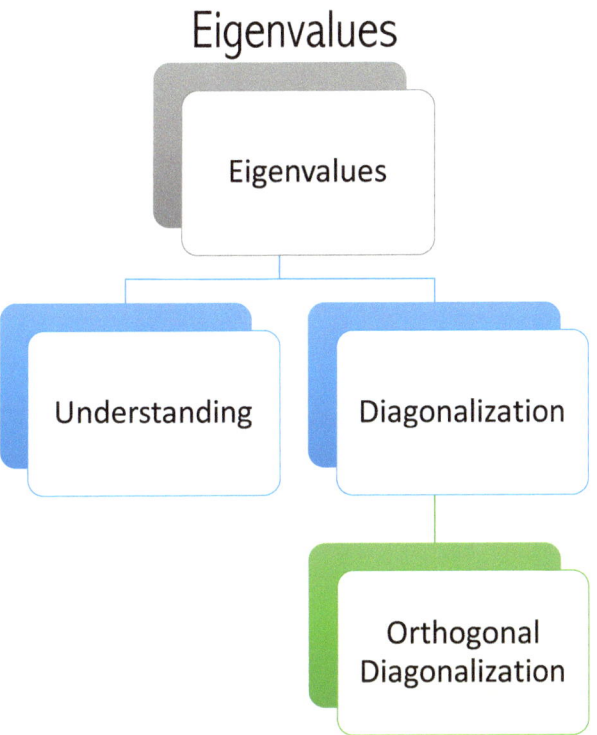

Definition Understanding:

$$Ax = \lambda x$$

λ: eigenvalue, a scalar

x: eigenvector corresponding to λ

$(A - \lambda I)x = 0$: contains nontrivial solution

$\det(A - \lambda I) = 0$ → characteristic equation

*eigenvectors of *distinct* eigenvalues are linearly independent
*an eigenvalue can have several eigenvectors
*an eigenvector can only belong to one eigenvalue
*each eigenvalue has its own eigenspace constructed by its eigenvectors

Geometric Understanding:

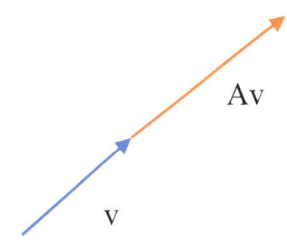

A acts on v to dilate/stretch it

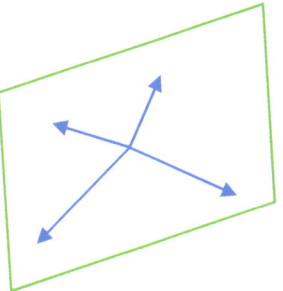

eigenspace for λ

Diagonalization

- There exists n linearly independent eigenvectors
- There are enough eigenvectors to construct an eigenvector basis
- $A = PDP^{-1}$
  P: invertible, contains eigenvectors
  D: eigenvalues along main diagonal

→ matrix A is diagonalizable

Orthogonal Diagonalization

- $A^T = A$
  symmetric matrix
- eigenvectors from different eigenspaces are orthogonal
- $A = UDU^T$
  or
  $A = UDU^{-1}$
  U: orthonormal

→ matrix A is orthogonally diagonalizable

Orthogonality

u and v are orthogonal vectors → u · v = 0
orthogonal complement- orthogonal to all vectors in a specified subspace
ex: $(\text{Row } A)^\perp = \text{Nul } A$ and $(\text{Col } A) = \text{Nul } A^T$

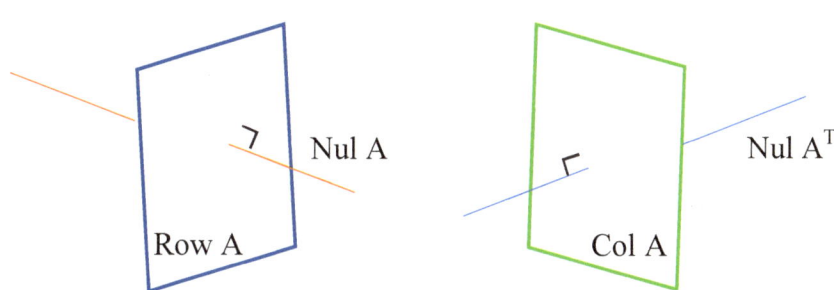

orthogonal set- all vectors are mutually perpendicular and linearly independent
orthogonal basis- basis for subspace spanned by the set

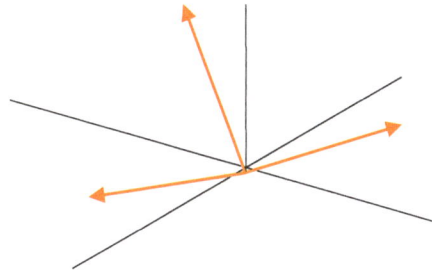

orthogonal projections

onto a line (u) → decompose y into 2 vectors
$\hat{y}$: multiple of u   z: orthogonal to $\hat{y}$
$y = \hat{y} + z$
if $\{u_1 \ldots u_p\}$ is an orthogonal basis for subspace W,
$c_j = \dfrac{y \cdot u_j}{u_j \cdot u_j}$ are the weights in the linear combination of $y = c_1 u_1 + \ldots + c_p u_p$

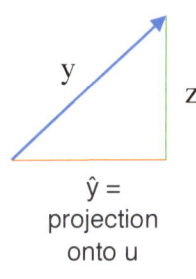
$\hat{y}$ = projection onto u

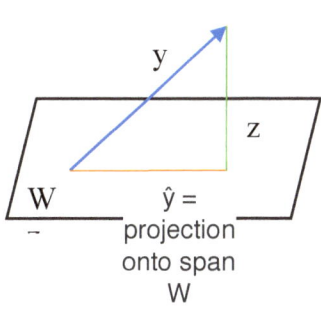
$\hat{y}$ = projection onto span W

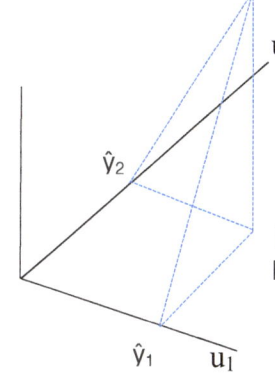
$\hat{y}$ = sum of its projections for each orthogonal basis vector

the Gram-Schmidt Process

- take normal vectors and make them orthogonal to each other
- useful for producing an orthogonal or orthonormal basis
- the span of the vectors before and after the manipulation is the same
- the Gram-Schmidt Process

given: $\{x_1, x_2, x_3\}$

$$v_1 = x_1$$
$$v_2 = x_2 - \frac{x_2 \cdot v_1}{v_1 \cdot v_1} v_1$$
$$v_3 = x_3 - \frac{x_3 \cdot v_1}{v_1 \cdot v_1} v_1 - \frac{x_3 \cdot v_2}{v_2 \cdot v_2} v_2$$

output: $\{v_1, v_2, v_3\}$, a set of now orthogonal vectors

v_3 is the vertical component left after x_3's projection onto the span W of v_1 and v_2

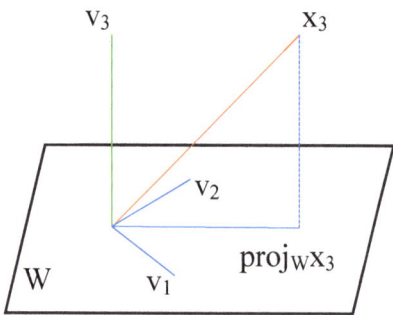

QR Factorization
$A = QR$

A: linearly independent columns
Q: columns form orthonormal basis for Col A (m x n)
R: upper triangular invertible matrix with positive entries along its diagonal (n x n)

$$Rx = Q^T b$$

Least Squares Solution
when to use → need a solution but none exists

- **best possible solution**
- **minimize the error**
- goal = finding an x that makes Ax as close as possible to b
- **projection of b onto Col A**

$$\hat{b} = \text{proj}_{\text{Col } A} b$$
$$A\hat{x} = \hat{b}$$

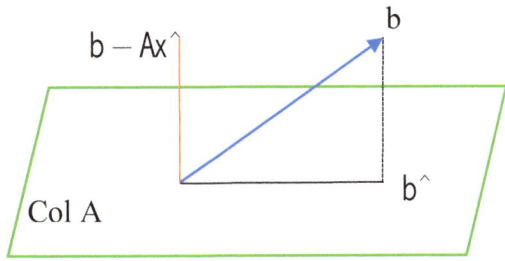

$b - A\hat{x}$ is orthogonal to each column of A: $a \cdot (b - A\hat{x}) = 0$
$$A^T(b - A\hat{x}) = 0$$
$$A^T b - A^T A \hat{x} = 0$$
$$A^T b = A^T A \hat{x}$$

Thm: these 3 statements are logically equivalent
1. $Ax = b$ has a *unique* least-squares solution for each b in R^m
2. columns of A are linearly independent
3. $A^T A$ is invertible

Linear Algebra Proof Bank
some common proofs

show A(u + v) = Au + Av

$$[a_1 \ a_2 \ a_3] \begin{bmatrix} u_1 + v_1 \\ u_2 + v_2 \\ u_3 + v_3 \end{bmatrix} = (u_1 + v_1)a_1 + (u_2 + v_2)a_2 + (u_3 + v_3)a_3$$

$$= (u_1 a_1 + u_2 a_2 + u_3 a_3) + (v_1 a_1 + v_2 a_2 + v_3 a_3)$$
$$= Au + Av$$

show A(cu) = c(Au)

$$[a_1 \ a_2 \ a_3] \begin{bmatrix} cu_1 \\ cu_2 \\ cu_3 \end{bmatrix} = (cu_1)a_1 + (cu_2)a_2 + (cu_3)a_3$$

$$= c(u_1 a_1) + c(u_2 a_2) + c(u_3 a_3)$$
$$= c(u_1 a_1 + u_2 a_2 + u_3 a_3)$$
$$= c(Au)$$

show (AB)$^{-1}$ = B^{-1}A^{-1}
　　just show → AB(B^{-1}A^{-1}) = AIA^{-1} = I

show (A^T)$^{-1}$ = (A^{-1})^T
　　just show → A^T is invertible and its inverse is (A^{-1})^T
　　1. (A^{-1})^T A^T = (AA^{-1})^T = I^T = I
　　2. A^T(A^{-1})^T = (A^{-1}A)^T = I^T = I

show Nul A is a subspace of R^n
　　just show → satisfies 3 conditions of a subspace
　　1. contains zero vector: Au = 0 and Av = 0
　　2. closed under vector addition: A(u +v) = Au +Av = 0+0 = 0
　　3. closed under scalar multiplication: A(cu) = cAu = c(0) = 0

prove coordinate mapping is a one-to-one linear transformation

for 2 vectors u and w: with same basis $\mathcal{B} = \{b_1, \ldots, b_p\}$

$$u = c_1 b_1 + \ldots + c_n b_n$$
$$w = d_1 b_1 + \ldots + d_n b_n$$
$$u + w = (c_1 + d_1)b_1 + \ldots + (c_n + d_n)b_n$$

$$[u+w]_\mathcal{B} = \begin{bmatrix} c_1 + d_1 \\ \vdots \\ c_n + d_n \end{bmatrix} = \begin{bmatrix} c_1 \\ \vdots \\ c_n \end{bmatrix} + \begin{bmatrix} d_1 \\ \vdots \\ d_n \end{bmatrix} = [u]_\mathcal{B} + [w]_\mathcal{B}$$

$$[ru]_\mathcal{B} = \begin{bmatrix} rc_1 \\ \vdots \\ rc_n \end{bmatrix} = r\begin{bmatrix} c_1 \\ \vdots \\ c_n \end{bmatrix} = r[u]_\mathcal{B}$$

one-to-one since $[u]_\mathcal{B} = [w]_\mathcal{B}$ → $\begin{bmatrix} c_1 \\ \vdots \\ c_n \end{bmatrix} = \begin{bmatrix} d_1 \\ \vdots \\ d_n \end{bmatrix}$ → u = w

prove eigenvectors corresponding to distinct eigenvectors are linearly independent

proof by contradiction → for now assume $\{v_1 \ldots v_p\}$ is dependent

$$c_1 v_1 + \ldots + c_p v_p = v_{p+1}$$
$$c_1 A v_1 + \ldots + c_p A v_p = A v_{p+1}$$
1. $c_1 \lambda_1 v_1 + \ldots + c_p \lambda_p v_p = A \lambda_{p+1} v_{p+1}$

$$(c_1 v_1 + \ldots + c_p v_p = v_{p+1}) \lambda_{p+1}$$
2. $c_1 \lambda_{p+1} v_1 + \ldots + c_p \lambda_{p+1} v_p = \lambda_{p+1} v_{p+1}$

1 – 2: $c_1(\lambda_1 - \lambda_{p+1})v_1 + \ldots + c_p(\lambda_p - \lambda_{p+1})v_p = 0$

because eigenvalues are distinct, none of $\lambda_1 - \lambda_{p+1} = 0$
hence to make this statement true, $\{v_1 \ldots v_p\}$ is independent

show similar matrices have same eigenvalues

similar matrices A and B: $B = P^{-1}AP$
same characteristic polynomial → same eigenvalues

$$B - \lambda I = P^{-1}AP - \lambda I$$
$$= P^{-1}AP - \lambda P^{-1}P = P^{-1}(AP - \lambda P)$$
$$= P^{-1}(A - \lambda I)P$$
$$\det(B - \lambda I) = \det(P^{-1})\det(A - \lambda I)\det(P)$$
$$\det(B - \lambda I) = \det(A - \lambda I)$$

note: similar matrices don't always have the same eigenvectors

show if A is symmetric ($A^T=A$) → eigenvectors from distinct eigenspaces are orthogonal

$$\lambda_1 v_1 \cdot v_2 = (\lambda_1 v_1)^T v_2 = (A v_1)^T v_2$$
$$= (v_1^T A^T) v_2$$
$$= v_1^T (A v_2)$$
$$= v_1^T (\lambda_2 v_2)$$
$$= \lambda_2 (v_1^T v_2)$$
$$\lambda_1 v_1 \cdot v_2 = \lambda_2 (v_1 \cdot v_2)$$
$$\lambda_1 v_1 \cdot v_2 - \lambda_2 (v_1 \cdot v_2) = 0$$
$$(\lambda_1 - \lambda_2)(v_1 \cdot v_2) = 0$$

eigenvalues are distinct → $\lambda_1 - \lambda_2 \neq 0$, thus $v_1 \cdot v_2 = 0$

show y decomposition into y and ŷ is unique

$$y = y_1\hat{\ } + z_1 \quad y_1\hat{\ } \text{ in } W \text{ and } z_1 \text{ in } W^\perp$$
$$\text{then} \quad \hat{y} + z = y_1\hat{\ } + z_1$$
$$\hat{y} - y_1\hat{\ } = z_1 - z$$

vector $v = \hat{y} - y_1\hat{\ }$ is in W and $W^\perp$
hence, $v \cdot v = 0 \rightarrow \hat{y} = y_1\hat{\ }$ and $z_1 = z$

note: orthogonal projection depends only on W and not on basis used

show $\|y - \hat{y}\| < \|y - v\|$ when v is any vector other than ŷ in span W

$$y - v = (y - \hat{y}) + (\hat{y} - v)$$
$$\|y - v\|^2 = \|y - \hat{y}\|^2 + \|\hat{y} - v\|^2$$
$$\|\hat{y} - v\|^2 > 0 \quad \text{bc } \hat{y} \neq v$$
$$\text{thus, } \|y - v\|^2 - \|y - \hat{y}\|^2 > 0$$

show the inverse matrix has same eigenvectors but reciprocal eigenvalues

$$Ax = \lambda x$$
$$x = A^{-1}\lambda x = \lambda A^{-1} x$$
$$\frac{1}{\lambda} x = A^{-1} x$$

Some True Statements to Keep in Mind

every matrix is row equivalent to a unique matrix in reduced echelon form

$A = PDP^{-1}$
$A^k = PD^kP^{-1}$

if a system has 2 different solutions → it must have infinite solutions

if A contains a row or column of zeros, then 0 is an eigenvalue of A

an nxn real symmetric matrix has n linearly independent eigenvectors

$(rA)^{-1} = r^{-1}A^{-1}$
$\det A^3 = (\det A)^3$
$(\det A)(\det A^{-1}) = 1$

every equation $Ax = 0$ has the trivial solution

row equivalent matrices have the same # of pivot columns

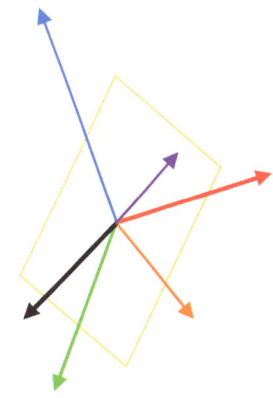

if $Ax = b$ has more than one solution → then so does $Ax = 0$

A and A^T have same characteristic polynomial → have same eigenvalues

show if A can be diagonalized, it's symmetric
$A^T = (UDU^T)^T$
$A^T = UD^TU^T = A$

one-to-one mapping → pivot in every column and every row

a 6x5 matrix cannot map $R^5 \to R^6$ bc not enough columns to have a pivot in every row

if A is diagonalizable, each vector in R^n can be written as a linear combination of the eigenvectors of A

if 2 vectors are in the null space of a matrix, then any linear combination of these 2 is also in the null space

if A and B are invertible, then AB is similar to BA
$AB = AB(AA^{-1}) = A(BA)A^{-1}$

if A and B are diagonal matrices, and $AB = C$, then $BA = C$

a diagonalizable matrix is not necessarily invertible → no direct relationship between diagonalizability and invertibility

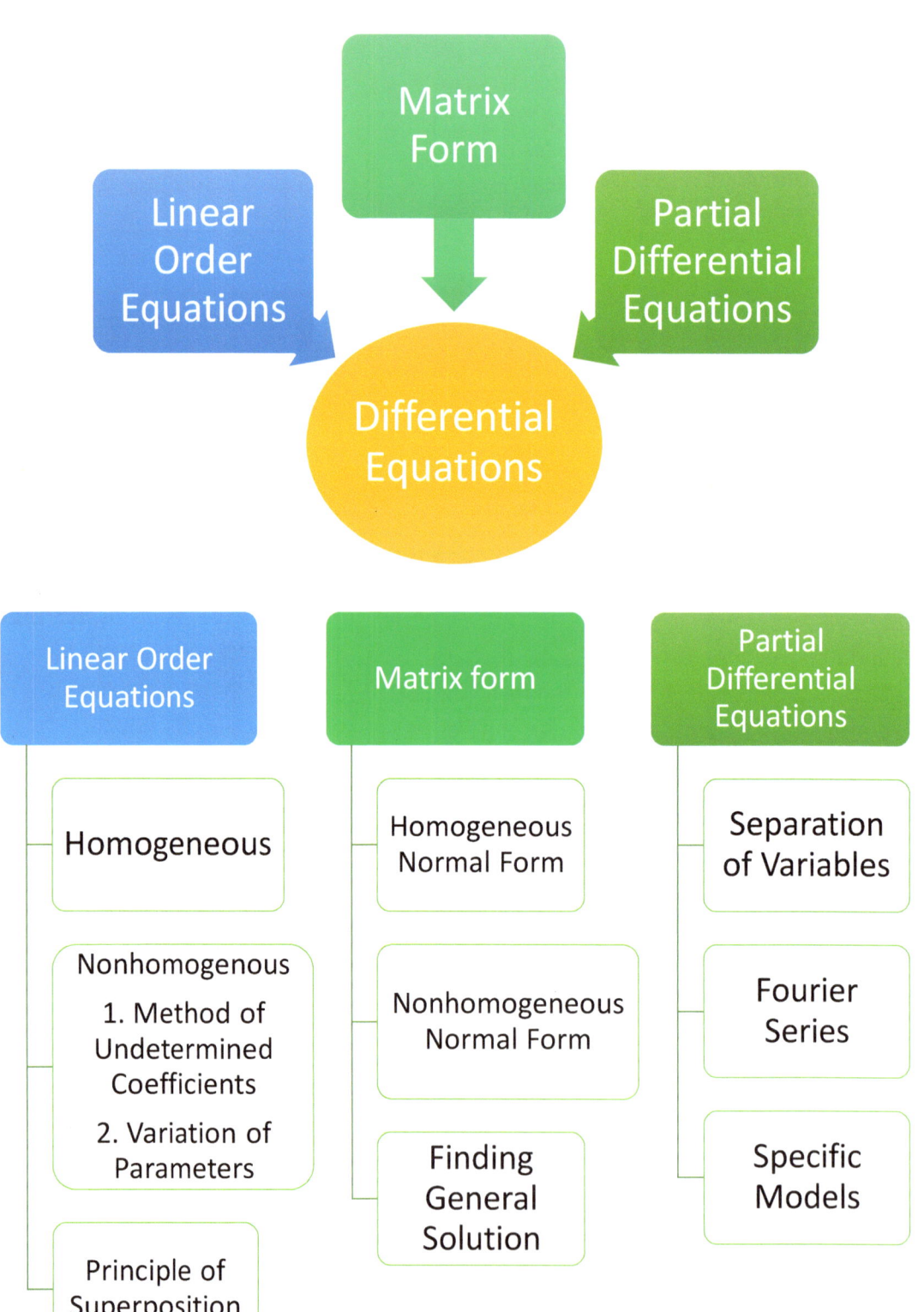

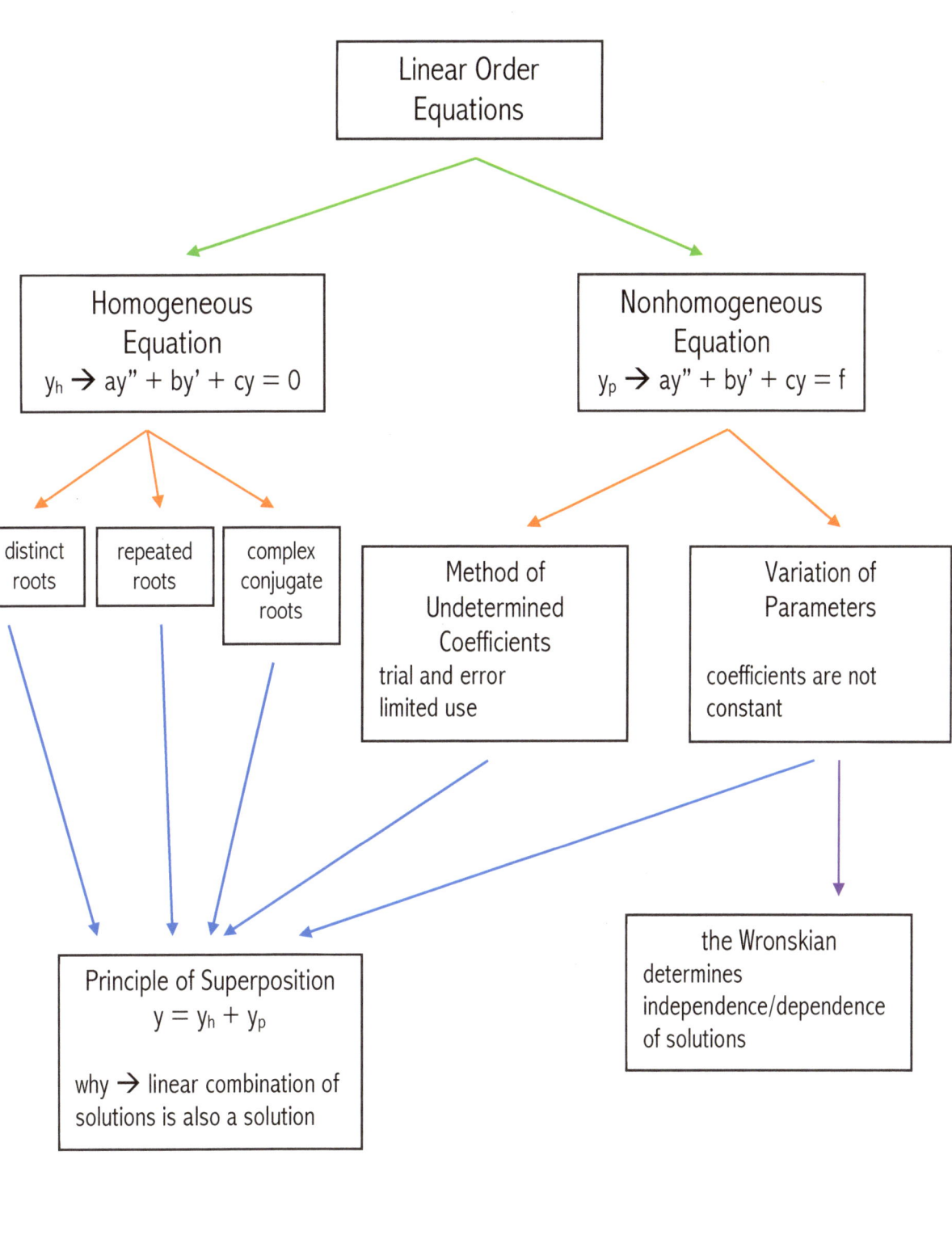

Physical Interpretation of a Linear Order Equation

The Mass Spring Oscillator

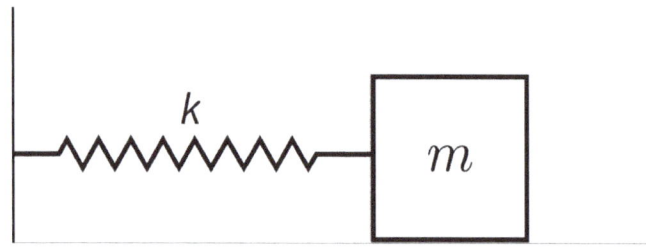

Newton's Second Law: $\qquad F = ma$

*acceleration is a second order differential equation of position with respect to time

$$a = \frac{d^2y}{dt^2} = y''$$

Hooke's Law: $\qquad F = -ky$

Equate Equations: $\qquad ma = -ky$

$$m\frac{d^2y}{dt^2} + ky = 0$$

$$\boxed{my'' + ky = 0}$$

With Damping: $\qquad F_{friction} = -bv$

$$= -by' = -b\frac{dy}{dt}$$

Sum all Forces: $\qquad \boxed{F_{ext}(t) = my'' + by' + ky}$

$$F_{ext} = [\text{inertia}]y'' + [\text{damping}]y' + [\text{stiffness}]y$$

Solving the Homogeneous Equation

$$ay'' + by' + cy = 0$$
$y = e^{rt}$ → auxiliary equation: $ar^2 + br + c = 0$

distinct roots: $y(t) = c_1 e^{r_1 t} + c_2 e^{r_2 t}$
repeated roots: $y(t) = c_1 e^{rt} + c_2 t e^{rt}$

complex conjugate roots: $\alpha \pm i\beta$
$$y(t) = c_1 e^{\alpha t} \cos \beta t + c_2 e^{\alpha t} \sin \beta t$$

*if initial conditions provided → substitute into found equation to find values of the unique coefficients

Solving the Nonhomogeneous Equation
$$ay'' + by' + cy = f$$
step 1: always find homogenous solution first
step 2: strategize which method to use to find particular solution

Method of Undetermined Coefficients
*applies only to polynomials, exponentials, sines, cosines, or product of these functions

$$ay'' + by' + cy = Ct^m e^{rt}$$
$$y_p(t) = t^s (A_m t_m + \ldots + A_1 t + A_0) e^{rt}$$

$s = 0$ → r is not a root
$s = 1$ → r is a simple root
$s = 2$ → r is a double root

$$ay'' + by' + cy = Ct^m e^{\alpha t} \cos\beta t \text{ or } Ct^m e^{\alpha t} \sin\beta t$$
$$y_p(t) = t^s (A_m t_m + \ldots + A_1 t + A_0) e^{\alpha t} \cos\beta t + t^s (B_m t_m + \ldots + B_1 t + B_0) e^{\alpha t} \sin\beta t$$

$s = 0$ → $\alpha + i\beta$ is not a root
$s = 1$ → $\alpha + i\beta$ is a root

Variation of Parameters

*now coefficients are functions of t aka not constant
*y_1 and y_2 are linearly independent solutions of the homogeneous equation

$$y_p(t) = v_1(t)y_1(t) + v_2(t)y_2(t)$$

system of equations:
$$v_1'y_1 + v_2'y_2 = 0$$
$$v_1'y_1' + v_2'y_2' = \frac{f}{a}$$

→ an imposed requirement to simplify calculations

$$v_1(t) = \int \frac{-fy_2}{aW} dt \qquad v_2(t) = \int \frac{fy_1}{aW} dt$$

the Wronskian (W)
*determinant containing n solutions and their derivatives
*square matrix = n-1 times differentiable

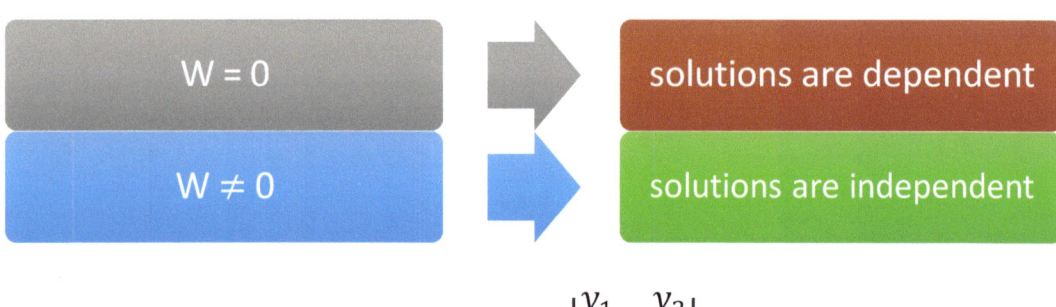

$$W = \det \begin{vmatrix} y_1 & y_2 \\ y_1' & y_2' \end{vmatrix}$$

Principle of Superposition

homogenous solution + particular solution = general solution
why? → any <u>linear combination</u> of solutions is also a solution

Matrix Form

Homogeneous Normal Form
$x' = Ax$

Nonhomogeneous Normal Form
$x' = Ax + f$

x' → vector of derivatives
A → coefficient matrix
x → solution vector

Finding General Solution
$x(t) = X(t)C$

Finding General Solution
$x(t) = X(t)C + x_p$

x_p → particular solution

C → coefficients
$X(t)$ → fundamental matrix, columns are solution vectors

1. find eigenvalues
2. find eigenvectors (u)

$$x(t) = c_1 e^{\lambda_1 t} u_1 + \cdots + c_n e^{\lambda_n t} u_n$$

Method of Undetermined Coefficients
$x' = Ax + tg$
$x_p = ta + b$
→ $a = A(ta + b) + tg$

system:
$Aa = -g \quad Ab = a$

for complex λ's: $\alpha \pm i\beta$ complex u's: $a \pm ib$

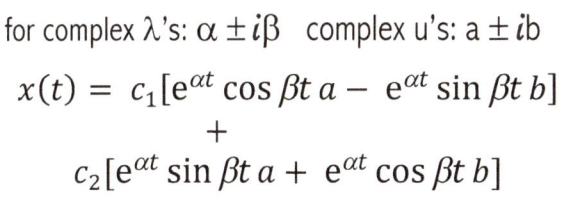

Variation of Parameters
$x_p(t) = X(t)v(t)$
$= X(t) \int X^{-1}(t) f(t) dt$

Understanding the Relationship Between Constant Coefficient Linear Order Equation and Matrix Form

Part 1: Linear Order Equation

$$ay'' + by' + cy = 0$$
$$ar^2 + br + c = 0$$

using quadratic equation → $r = \dfrac{-b \pm \sqrt{b^2 - 4ac}}{2a}$

Part 2: Converting Linear Order Equation into Matrix Form

$$\boxed{\begin{array}{ll} x_1 = y & x_2 = y' \\ x_1' = x_2 & x_2' = y'' \end{array}}$$

$$x_1' = x_2$$

substitute into linear equation → $ax_2' + bx_2 + cx_1 = 0$

$$x_2' = \dfrac{-c}{a} x_1 + \dfrac{-b}{a} x_2$$

$$\begin{bmatrix} x_1' \\ x_2' \end{bmatrix} = \begin{bmatrix} 0 & 1 \\ \dfrac{-c}{a} & \dfrac{-b}{a} \end{bmatrix} \begin{bmatrix} x_1 \\ x_2 \end{bmatrix}$$

find eigenvalues of matrix A

$$\det \begin{bmatrix} 0 - \lambda & 1 \\ \dfrac{-c}{a} & \dfrac{-b}{a} - \lambda \end{bmatrix}$$

$$(-\lambda)\left(\dfrac{-b}{a} - \lambda\right) + \dfrac{c}{a} = 0$$
$$a\lambda^2 + b\lambda + c = 0$$

using quadratic equation → $\lambda = \dfrac{-b \pm \sqrt{b^2 - 4ac}}{2a}$

Part 3: Compare

eigenvalues of roots to linear equation of Part 1 are the same as the eigenvalues of matrix A in Part 2

Partial Differential Equations

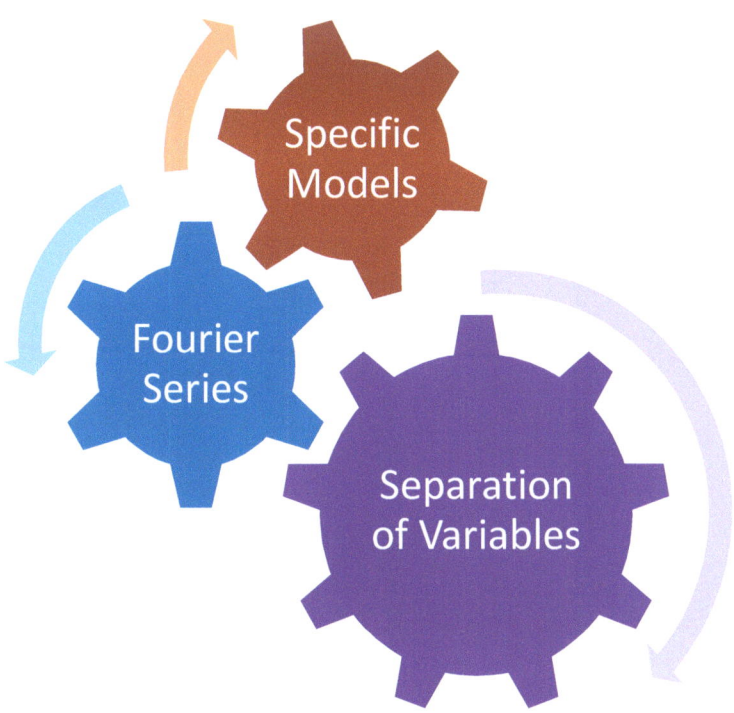

Separation of Variables
Goal: reduce a partial differential equation into 2 ordinary differential equations

$$u(x, t) = X(x)T(t)$$
$$\frac{\partial u}{\partial t}(x, t) = X(x)T'(t) \qquad \frac{\partial^2 u}{\partial t^2}(x, t) = X''(x)T(t)$$

substitute into heat flow equation:
$$X(x)T'(t) = \beta X''(x)T(t)$$

separate variables into 2 eq. by setting ratios equal to same constant:
$$\frac{X''(x)}{X(x)} = -\lambda \quad \frac{T'(t)}{\beta T(t)} = -\lambda$$

finding eigenfunctions: values of λ for which nontrivial solutions exist
- case 1: $\lambda < 0$ → roots: $\pm\sqrt{-\lambda}$
- case 2: $\lambda = 0$ → repeated root: $r = 0$
- case 3: $\lambda > 0$ → roots: $\pm i\sqrt{\lambda}$

Fourier Series

main idea: modeling a composite complex function as a sum (linear combination) of many simpler sinusoidal waves
basically: the sum of sines and cosines forms a complex function
*more terms of sines and cosines → closer to actual function f(x) you get

$$f(x) \sim \frac{a_0}{2} + \sum_{n=1}^{\infty} \left(a_n \cos \frac{n\pi x}{L} + b_n \sin \frac{n\pi x}{L} \right)$$

$$a_0 = \int_{-L}^{L} f(x) dx$$

$$a_n = \frac{1}{L} \int_{-L}^{L} f(x) \cos \frac{n\pi x}{L} dx \qquad b_n = \frac{1}{L} \int_{-L}^{L} f(x) \sin \frac{n\pi x}{L} dx$$

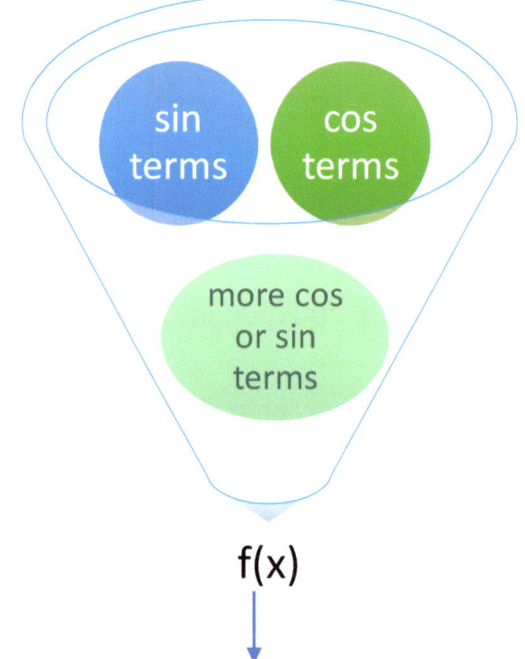

half range expansion: extend f(x) to turn it into an odd or even function
odd: f(x) = f(x) {0 < x < L}   even: f(x) = f(x) {0 < x < L}
f(-x) = -f(x) {−L < x < 0}   f(-x) = f(x) {−L < x < 0}

Fourier Sine Series
if f(x) is odd → cos terms disappear

$$f(x) \sim \frac{a_0}{2} + \sum_{n=1}^{\infty} \left(b_n \sin \frac{n\pi x}{L} \right)$$

$$b_n = \frac{2}{L} \int_{0}^{L} f(x) \sin \frac{n\pi x}{L} dx$$

Fourier Cosine Series
if f(x) is even → sin terms disappear

$$f(x) \sim \frac{a_0}{2} + \sum_{n=1}^{\infty} \left(a_n \cos \frac{n\pi x}{L} \right)$$

$$a_n = \frac{2}{L} \int_{0}^{L} f(x) \cos \frac{n\pi x}{L} dx$$

Specific Models

The Heat Flow Equation

model: measuring heat flow through rod of length L

u(temp of wire) depends on t (time) and x (position within wire)

$$\frac{\partial u(x,t)}{\partial t} = \beta \frac{\partial^2 u(x,t)}{\partial x^2}$$

$u(0, t) = u(L, t) = 0$: boundary conditions (temp at ends of rod is 0°C)

$u(x, 0) = f(x)$: initial temperature distribution function

formal solution:

$$u(x,t) = \sum_{n=1}^{\infty} \left(c_n \, e^{-\beta \left(\frac{n\pi}{L}\right)^2 t} \sin \frac{n\pi x}{L} \right)$$

The Wave Equation

model: vibrating string of length L with ends held fixed

$u(x, t)$ = displacement of string

$$\frac{\partial u(x,t)}{\partial t} = \alpha^2 \frac{\partial^2 u(x,t)}{\partial x^2}$$

$$u(0, t) = u(L, t) = 0$$

$u(x, 0) = f(x)$: initial displacement

$$\frac{\partial u(x,t)}{\partial t} = g(x) : \text{initial velocity}$$

$$u(x, t) = \sum_{n=1}^{\infty} \left(a_n \cos \frac{n\pi x}{L} + b_n \sin \frac{n\pi x}{L} \right) \sin \frac{n\pi x}{L}$$

$$f(x) = \sum_{n=1}^{\infty} \left(a_n \sin \frac{n\pi x}{L} \right)$$

$$g(x) = \sum_{n=1}^{\infty} \left(b_n \frac{n\pi x}{L} \sin \frac{n\pi x}{L} \right)$$

A Return to What We First Saw:
Understanding the Relationship Between Linear Algebra and Differential Equations

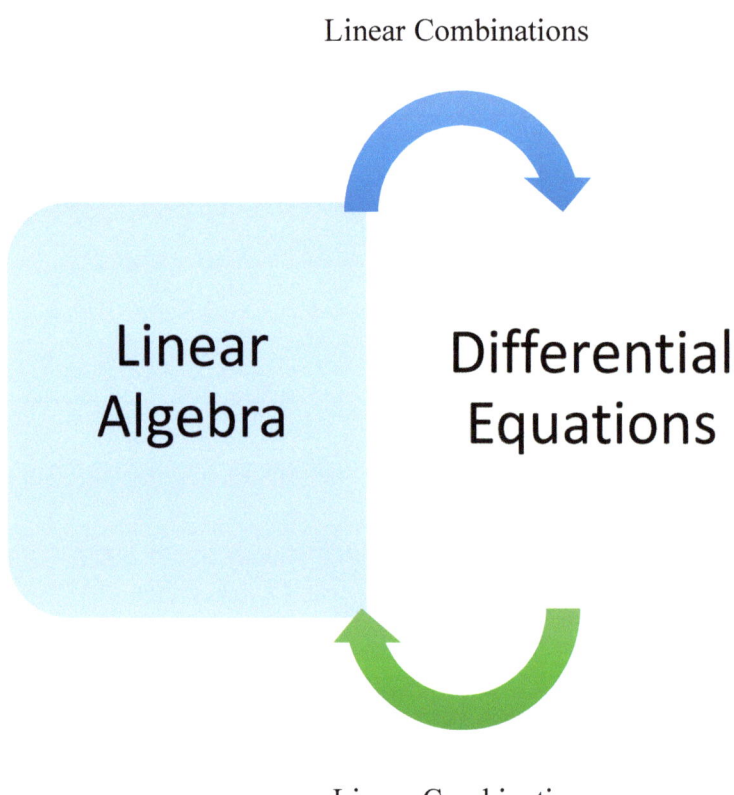

the idea of Linear Combinations links Linear Algebra and Differential Equations

Hopefully, this page makes much more sense at the conclusion of this book.

www.ingramcontent.com/pod-product-compliance
Lightning Source LLC
Chambersburg PA
CBHW041300180526
45172CB00003B/909